BACKPACKER.

Predicting
Weather

FORECASTING, PLANNING, AND PREPARING

Lisa Densmore

FALCONGUIDES

GUILFORD, CONNECTICUT
HELENA, MONTANA

AN IMPRINT OF GLOBE PEQUOT PRESS

To Jack, who chases all of my storm clouds away

To buy books in quantity for corporate use
or incentives, call **(800) 962–0973**
or e-mail **premiums@GlobePequot.com**.

FALCONGUIDES®

Backpacker is a registered trademark of Cruz Bay Publishing, Inc.
FalconGuides is an imprint of Globe Pequot Press.
Falcon, FalconGuides, and Outfit Your Mind are registered trademarks of Morris Book Publishing, LLC.

Photos on pages 10, 36, 37, 43, and 63 are courtesy of Jack Ballard. All others are by Lisa Densmore.
Text design by Sheryl P. Kober
Page layout by Melissa Evarts

Library of Congress Cataloging-in-Publication Data

Densmore, Lisa Feinberg.
 Backpacker magazine's predicting weather : forecasting, planning, and preparing / Lisa Densmore.
    p. cm.
 Includes index.
 ISBN 978-0-7627-5656-8 (alk. paper)
1. Weather forecasting—Popular works. 2. Weather—Popular works. 3. Backpacking. I. Backpacker. II. Title. III. Title: Predicting weather.
 QC981.2.D46 2010
 551.63—dc22

                                                                2009046681

Printed in China

10 9 8 7 6 5 4 3 2 1

Contents

Mount Washington in New Hampshire looks benign on a clear day but is home to some of the wildest weather in the world.

INTRODUCTION

The summit of Mount Washington in New Hampshire's Presidential Range is a modest 6,288 feet above sea level, yet this mountain continually surprises experienced backcountry travelers with some of the most extreme weather on the planet. The highest winds on the earth's surface, 231 miles per hour, were recorded there in April 1934. It's always windy and much colder than you expect atop Mount Washington. In fact, the highest temperature ever recorded on the summit is only 72°F (August 1975). On a typical summer day, when it's 85° at the trailhead, it's usually a nippy 50° or cooler at the top, not including the wind, which averages 35 miles per hour.

Seasoned Mount Washington hikers prepare for snow even in July. And though I know this and always pack a hat, gloves, and extra clothing whenever I venture into the Presidential Range, the weather still throws me a curveball on occasion. The last time I climbed Mount Washington, the wind gusted to over 70 miles per hour; I could hardly stand up. Each time I planned a step, the wind shoved my hiking boot 10 inches to the right, which made negotiating the rocks rather hazardous. Even more treacherous than a misstep in the talus was the risk of hypothermia. Despite layers of clothing under my shell, the wind threatened to pull every ounce of heat from my body. I made it to the Appalachian Mountain Club (AMC) Lakes of the

Clouds Hut at the base of the summit cone, drained a mug of hot chocolate, then buried myself in my sleeping bag and three heavy wool blankets. I awoke two hours later, exhausted but warm again.

Exposure is one of the leading causes of death among backcountry travelers. It has a single cause: being outside in weather for which you are unprepared. Even if conditions are not life-threatening, they can still be miserable. In town, getting a weather forecast is often easier than finding a glass of water, but the farther you get from civilization, the less likely it is that you'll be able to pick up the forecast on a radio, a cell phone, or any other electronic device. What's more, the forecast in town is likely much milder than the weather in the woods.

Weather can change quickly, too. There's an old saying in the Presidential Range: "If you don't like the weather, wait a moment." Funny how that saying also crops up in the southern Appalachians, the Rockies, the Sierra Nevadas, and most any other mountain range! The key is determining what threat a change in the weather might pose so you can respond appropriately. Every mountain range is susceptible to lightning strikes, which are particularly dangerous if you are above the tree line. Or a snowstorm might pass through the area even in the middle of summer.

Then there's desert backcountry, where it's hot, hotter, and hottest, contrasted with cold at night. But how hot or cold? The difference between hiking on an 85º day and a 105º day can mean the difference between serious perspiration and serious heatstroke.

Weather forecasting in the backcountry provides more than comfort and campfire talk—it can be a matter of life and death. This book will help you understand how weather works, and more important, it will give you key weather-predicting skills to help make your trips into the backcountry, whether with a pack, a saddle, or a paddle, more enjoyable and safer.

Knowing the weather is key to planning your gear and your route in the backcountry.

Chapter One:
Weather Basics

weath·er *n.* 1. The state of the atmosphere at a given time and place, described by specification of variables such as temperature, moisture, wind velocity and pressure.

—*The American Heritage Dictionary of the English Language*

Weather surrounds you. It is simply the condition of the air and how it's flowing. But it's always changing. Sometimes that change is very, very slow, and sometimes it comes with blinding speed and drenching, staggering force. Various aspects of our world—some local to where you might be backpacking, such as a massive peak, and others distant, such as an ocean— affect the weather and the rate at which weather changes. Basic forces of nature, such as gravity and sunlight, also affect air movement, both cyclically and in relation to a specific weather event.

The atmosphere moves constantly and in more than one direction. Air near the ground warms up during the day, pulling moisture in the form of water vapor into the atmosphere as it rises. As moisture gains elevation, it cools, condensing around dust particles and forming clouds. When it cools to its dew point, it falls back to the earth as precipitation.

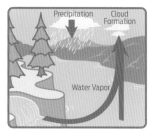

The evaporation-precipitation cycle

How simple weather forecasting would be if it were just based on hot air rising, but in addition to moving up and down, air also moves across the land as wind. In the backcountry, we are acutely aware of wind. We feel it as a refreshing breeze, a gusty gust, or a numbing chill. And then there are those nasty overbearing winds that cause your tent to shake and rattle, keeping you up all night.

As if things aren't complicated enough with air moving both vertically and horizontally at the same time, the temperature of the air, the speed of the air, and the atmospheric pressure also contribute to the weather around you and, more important, to what's coming. Weather forecasting is complex. The National Oceanic and Atmospheric Administration (NOAA) recently completed the installation of a $180 million computer system, which can make 69.7 trillion calculations per second, to more accurately predict weather. Even with this state-of-the-art computer, weather forecasting, at least long-range forecasting, remains inexact. You can accurately predict the

Backpackers approaching Mount Wood in the Absaroka-Beartooth Wilderness in Montana on a clear day.

weather over the next twenty-four to forty-eight hours, and you don't need a computer to do it. You only need to observe and interpret the air around you.

When the weather feels calm, air molecules still move but slowly and in unison, therefore making the weather more predictable. On the other hand, when a mass of cold air collides with a mass of warm air and the amount of moisture in each air mass is different (which is usually the case), the air molecules move in different directions and speeds depending on how high they are above the ground. The faster the air

mass travels, the more violent the collision and thus the weather. The slower the air mass, the more gradual the change, and if that change is toward bad weather, the longer the unpleasant conditions will linger.

FOUR INGREDIENTS OF WEATHER

The recipe for weather contains only four ingredients:

1. Air temperature
2. Wind speed
3. Humidity
4. Barometric pressure

When you mix these ingredients together, not only do you get the atmospheric conditions at a particular moment but you can also predict what's coming in the next twenty-four to forty-eight hours with a high degree of accuracy.

Air Temperature. Air temperature affects you in the backcountry in two ways. First, the temperature of the air around you near the ground determines how many layers you need to wear in order to stay comfortable. Second, the temperature of the air high above you determines whether you wear or carry your raingear. As warm air rises and cools off, it gets closer and closer to its dew point, which is the temperature at which water vapor in the air turns to water droplets. If the droplets become heavy enough, they fall back to earth as precipitation.

Wind Speed. Like air temperature, wind speed affects you in the backcountry in two ways. First, the perceived temperature on exposed skin is lower than the actual air temperature due to the wind. The

Low scud clouds under cirrus clouds means the sun may shine for a moment, but the rain will return soon.

stronger the wind, the colder the air feels. Second, wind is a symptom of changing weather. A strong wind can mean a cold front is approaching, with the possibility of severe thunderstorms on its leading edge.

Humidity. Humidity is the amount of water vapor in the air, though "relative humidity" is more important when it comes to predicting weather because it is related to the dew point. When the air reaches its dew point, it cannot hold any more water vapor and

must release it. As air rises, it cools off, getting closer and closer to its dew point. In addition, as air rises, relative humidity increases. A relative humidity of 100 percent means the air temperature equals the dew point, thus the air is saturated with water. Be prepared for a wet day outdoors if the relative humidity is high.

Barometric Pressure. Barometers measure air pressure. The air in a warm air mass is always lighter (less dense) than the air in a cold air mass; thus warm air exerts less pressure on the earth than cold air. If the barometric pressure is falling, that means a warm front is coming in. If the barometric pressure is rising, a cold front is arriving. If the barometric pressure is steady, the weather is unlikely to change until the air pressure begins to move up or down.

Chapter Two:
Reading the Sky

Want to know the weather when you crawl out of your tent tomorrow? Look at the clouds today! The types of clouds in the sky, along with the speed and direction they are moving, are helpful forecasters.

TYPES OF CLOUDS

There are four families of clouds: cirrus, cumulus, stratus, and lenticular, though lenticular clouds are often

Backpacker in the rain

lumped in with stratus clouds. Within these major cloud families, there are several subfamilies based on altitude, size, and color. In addition, there are combinations of clouds. Here's how to identify what's in the sky above you and what weather to expect:

Cirrus

Cirrus clouds form high in the sky, at elevations over 20,000 feet. Sometimes called mare's tails, they look like wisps and swirls. Though cirrus clouds do not release precipitation, they typically signal an incoming warm front, which means worsening weather, though it might take twelve to twenty-four hours to arrive. Note: Cirrus clouds that result from the condensation trails from jet airplanes do not signal an incoming storm.

Cirrus clouds

Cumulus

Cumulus clouds, sometimes called "heap clouds," look like piles of cotton that build upward in the sky. The bottom elevation of the cloud depends on the amount of moisture in the air. In humid climates the base can be as low as 6,500 feet. In arid climates and in the mountains, it might be as high as 20,000 feet. If the top of a cumulus cloud becomes cold enough, it reaches its dew point and precipitation falls from it. If the ground temperature is above freezing, expect rain. If the ground temperature is below freezing, expect snow. While

Cumulus clouds

snow sometimes falls when ground temperatures are slightly above freezing, ground temperature is a good gauge for determining what form precipitation will take. First, it's available. You can carry a thermometer. Second, it's almost always below freezing aloft, especially at the top of tall cumulus clouds, which can grow to elevations above 40,000 feet. But the snowflakes that form high in the sky usually turn to raindrops by the time they reach you on the trail if the thermometer on your pack reads above 32 degrees.

Stratus

Stratus means "layered," although stratus clouds have no discernable form. They can form at zero feet as fog or at 20,000-plus feet. Stratus clouds turn the sky a featureless sheet of white or gray, often blocking out the sun. It's a gloomy day under stratus clouds, though you might escape precipitation. If rain does come, it will likely fall in the form of drizzle.

Stratus clouds

Lenticular

Lenticular clouds are elongated clouds caused by wind as it passes over the top of a mountain. Air is forced upward as it hits the side of the mountain, then curls over the peak. A lenticular cloud forms on the leeward side of a summit at the "wave break." These clouds look stationary, but they signal very strong wind.

Lenticular clouds

COMMON COMBINATION CLOUDS

Cirrostratus: Cirrus clouds forming a sheet across the sky at a very high elevation.
Forecast: Approaching warm front. Precipitation in twelve to twenty-four hours.

Cirrostratus clouds

Cirrostratus with halo: An intensifying sheet of very high clouds, barely discernible at first except for a halo around the sun or moon.

Forecast: Cloud ceiling is lowering. Possibility of precipitation within forty-eight hours.

Cirrostratus clouds with halo

Cirrocumulus: Very high clouds that look like a mass of puffballs or fish scales rather than wisps.

Forecast: Fair but cold weather now. Precipitation may be on its way due to an unstable air mass. In tropical regions cirrocumulus clouds may indicate an impending hurricane, thus the expression, "mackerel sky, storm is nigh."

Cirrocumulus clouds

Altocumulus: Similar to cirrocumulus clouds but lower (about 8,000 feet).

Forecast: Precipitation or thunderstorm within twenty-four hours. If you see altocumulus clouds in the morning, keep your rain gear handy. You'll likely need it after lunch.

Altocumulus clouds

Cumulonimbus: Massive cumulus clouds, often with anvil-shaped tops.

Forecast: Take cover! An intense downpour and possibly hail are imminent and probably accompanied by a violent thunderstorm, but it should be over in twenty minutes.

Cumulonimbus clouds

Cumulocongestus: The precursor to cumulonimbus clouds but still forming. The top is uneven and growing higher and higher.

Forecast: Unstable air mass. A thunderstorm is coming, but you've got a little more time than with cumulonimbus clouds.

Cumulocongestus clouds

Fair-weather cumulus: Puffy white clouds that form later in the day. They may form in lines across the sky, indicating the direction of the breeze. Birds may circle in the thermals under them.

Forecast: Slather on more sunscreen. The air mass is stable. Enjoy the rest of a beautiful day.

Fair-weather cumulus clouds

Altostratus: Medium-high clouds (8,000 feet) that turn the sky medium gray. You can see a bright spot where the sun is.

Forecast: Light precipitation may fall now. Expect rain within forty-eight hours if the sky gets darker.

Altostratus clouds

Nimbostratus: Low, dark, thick clouds with ragged edges and undefined shapes. Also called "scud clouds."

Forecast: Tent cribbage, anyone? It's going to rain steadily all day.

Nimbostratus clouds

MORE CLUES FROM THE CLOUDS

A savvy weather watcher can make fairly accurate predictions based on what the clouds are doing. Here are a few more clues to help you decide whether to wear your poncho or your sunglasses:

It's going to be wet today if . . .
>> the clouds build in size and quantity;
>> the speed of the clouds increases;
>> the higher clouds move in a different direction from that of the lower clouds;
>> low dark clouds scurry under high dark clouds.

It's going to be dry today if . . .
>> the sky is blue and there's no wind;
>> the sky is white with very high clouds and there's no wind;
>> the sky is dotted with rows of white, puffy clouds.

Chapter Three:
Figuring Out Fronts

Colliding air masses are known as fronts. In the northern hemisphere, the prevailing winds move from west to east. As a result, general weather patterns move from west to east. Like one car rear-ending another, the incoming front from the west rams into the outgoing front, pushing it eastward. Also as in a car collision, the faster and more aggressive the new front is, the more violent the collision and thus the stormier the weather it causes. That's why some changes in weather are merely fender benders and others cause irreparable wreckage.

There are three types of weather fronts: cold fronts, warm fronts, and occluded fronts. Each brings bad weather; it's just a matter of how bad and what comes after the front passes through.

WARM FRONTS

Symptoms: Low barometric pressure, high humidity, low cloud ceiling, poor visibility.

What's Really Happening: A warm air mass rises slowly above the cold air in front of it. As the warm air rises, it eventually cools to its dew point.

Resulting Weather: Relatively calm with maximum winds of 20 miles per hour at the leading edge of the front. Steady rain for several days.

Cloud Sequence: Cirrus, followed in succession by cirrostratus, altostratus, and finally nimbostratus rain clouds.

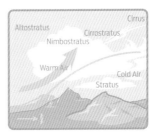

Warm front

COLD FRONTS

Symptoms: High barometric pressure, high cloud ceiling, good visibility unless precipitation is present.

What's Really Happening: Unstable, fast-moving cold air pushes under the warm air in front of it. The warm air mass is forced upward, which cools it. If it reaches its dew point, the resulting precipitation may be heavy and violent.

Resulting Weather: Fair weather, although it may change with little warning. Strong winds up to 35 miles per hour, generally from the north or west in the northern hemisphere. Severe but short-lived thunderstorms or heavy snow squalls.

Cloud Sequence: Altostratus, then nimbostratus or cumulonimbus rain clouds.

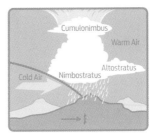

Cold front

WARM OCCLUDED FRONTS

Symptoms: Change of wind direction, usually from south-southeast to north-northwest; falling, then rising barometric pressure; poor visibility during precipitation, then improving.

What's Really Happening: A cold front overtakes a warm front, lifting (occluding) the warm air mass off the earth's surface. This allows the incoming cold front to collide with the cold front that's departing ahead of the lifted warm air mass. The incoming cold front is warmer than the departing cold front, climbing over it while keeping the warm front in the middle above both cold fronts.

Resulting Weather: Thunderstorms possible. Light to heavy rain followed by dry weather after the front moves out. Cold temperatures followed by slightly milder temperatures.

Cloud Sequence: Cirrus, cumulostratus, altostratus, nimbostratus, scattered cumulus.

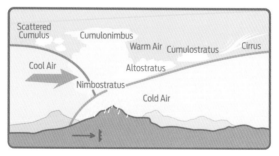

Warm occluded front

COLD OCCLUDED FRONTS

Symptoms: Change of wind direction, usually from south-southeast to north-northwest; falling, then rising barometric pressure; poor visibility during precipitation, then improving.

What's Really Happening: A cold front overtakes a warm front, lifting (occluding) the warm air mass off the earth's surface. This allows the incoming cold front to collide with the cold front that's departing ahead of the lifted warm air mass. The incoming cold front is colder than the departing cold front and wedges under it.

Resulting Weather: Thunderstorms possible; light to heavy rain, followed by drying weather. Funnel cloud in severe cases. Cold temperatures get even colder.

Cloud Sequence: Cirrus, cumulostratus, altostratus, nimbostratus, then scattered cumulus.

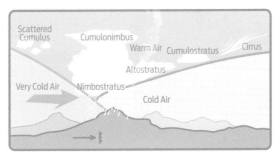

Cold occluded front

Chapter Four:
Dew, Frost, and Fog

While we usually look at the sky for clues to incoming weather, what's on the ground can be a helpful predictor, too. The three common weather-related ground phenomena are dew, frost, and fog, all of which affect camping and backcountry travel.

DEW

Waking in the morning to find everything you left outside your tent soaking wet is not an ideal way to start the day, particularly if you need to break camp before it dries out. Dew does not fall from the sky, so it is technically not precipitation, but it does come from water vapor in the air that has cooled below its condensation point. Dew forms when objects such as your tent, the

Dew on spruce

grass, or the backpack you left under a tree become colder than the air around them. Water vapor in the air then condenses on the object. The closer you camp to water, such as a lake or a river, the more likely it is that there will be dew on your gear in the morning. Dew tends to be heavier on a valley floor versus on a ridge, in part because there's usually a stream at the bottom of the valley, though if it's a clear night when radiant cooling is high, there will be dew on the ridge, too.

Dealing with Dew

Here are several ways to avoid having dew ruin your morning, especially if you have to break camp:

» Pick a campsite that is as high as possible. A small plateau or high rocky shelf above water will be slightly drier than a spot by water's edge. Put on the tent fly if the humidity is above 50 percent. Though it's lovely to lie in your tent and gaze at the Milky Way through the netting, you will wake up wet without the fly.

Use the tent fly to keep dew off you while you sleep.

Put cooking items on plastic to keep them dry.

» If your pack doesn't fit inside your tent or in the tent vestibule, put a rain cover over it or cover it with a large plastic garbage bag before you go to sleep.

» Put your cooking gear inside a garbage bag, or lay a tarp over it to keep it dry overnight. In the morning use a dry garbage bag or small tarp under your cooking gear to keep it off the wet ground while you prepare breakfast. Note: Never use your cookstove on the tarp!

» When purchasing a tent, check that its seams are sealed so dew doesn't seep through the needle holes.

» If you put a tarp under your tent, make sure the tarp does not extend past the edges of the tent; otherwise dew drops will collect under your tent floor, and you might wake up in a puddle.

» If it's legal in the area where you are camping, build a small morning campfire to help dry out your gear before you pack it up. Remember to cover your firewood before you go to sleep to keep it dry, too.

» Allow your tent to dry before packing it up. You might have to wait an extra hour before departing, but you'll extend the life of your tent, and it's a lot more pleasant to set up a dry tent at your next campsite.

Predicting Weather by Dew

Heavy dew early in the morning or late in the evening typically means fair weather for at least the next twelve hours, though all bets are off if it rained the night before—rain increases the amount of moisture at ground level. Likewise, if the ground is dry (dewless) in the morning, expect rain later that day.

FROST

Frost forms when water vapor comes in contact with an object that is at or below 32°F, condensing into ice crystals. There are two types of frost—depositional frost, also known as white frost or hoarfrost, and frozen dew. Depositional frost forms when water vapor skips its liquid form and goes directly to a solid. If thick enough, it resembles a coating of light snow. Frozen dew forms when dew droplets condense on an object, then turn to ice as the temperature drops below freezing. It does not have a crystal pattern and can be very slippery and hard to see.

Depositional frost on tent.

Dealing with Frost

As with dew, keep everything covered or inside plastic, and put the fly on your tent. (See "Dealing with Dew," on page 25.) Here are a few more suggestions to make breaking camp more comfortable on a frosty morning:

» It's difficult to pack up a stiff, frost-covered tent, so relax with another cup of coffee and wait for the day to warm up a little before trying to break down your campsite.

Morning campfires help dispel frost.

» Like black ice on the road, if frozen dew formed overnight, the rocks on the trail may be slick. Let the day warm up before hiking for friendlier footing.

» Wear gloves. If there's frost, temperatures remain chilly even when they rise above freezing. A pair of waterproof gloves or quick-drying fleece gloves will keep your hands warmer while handling metal tent rods and camping pots.

Predicting Weather by Frost

As with dew, a heavy frost in the morning or late evening usually means fair weather for at least twelve hours. Frozen dew in the morning is a sign that a cold front came through during the night. The high temperature for the day may not be high at all, but the sun will shine.

FOG

Fog is a stratus cloud that forms near the ground. When the surface of the earth cools, water vapor near the ground condenses into fog. When we talk about fog "burning off," it's really evaporating (becoming water vapor again) as the sun warms the air.

Fog and mist are similar; fog is merely thicker. Pilots pay serious attention to fog when visibility becomes less than half a mile, but it really doesn't

Fog

affect backcountry travelers in terms of navigation until it drops below 600 feet and becomes "thick as pea soup." When you can no longer see the trail markers, foggy conditions become dangerous because of the higher risk of misreading the terrain and falling, or simply getting lost. Both of these situations can be serious, even deadly, if you are alone or far from a trailhead.

By recognizing the conditions around you that lead to fog, you can better predict when the fog is going to lift and what the weather will be like when it does. It helps to recognize the different types of fog and know what to expect from each one.

Radiation Fog

» It forms on a clear windless night as the ground loses heat by radiant cooling.
» It is typically patchy and confined to low-lying areas but can become widespread.
» It burns off as the day heats up but can hang around during the winter.

Radiation fog

Advection Fog

» It forms when warm, moist air passes over cold ground, particularly snowpack or near a coastal shoreline.

» It begins three to four days after a front passes and stays until the next front comes through.

» Expect moderate winds up to 10 miles per hour.

Advection fog

Hill Fog or Upslope Fog

» It forms when mild, moist air travels up the windward side of a mountain range.
» It covers a large area on the lower part of mountains, where the land slopes gradually upward.
» It is more common during the winter.
» It can be dense, so stay around the trees for better visibility until you hike above it.

Hill fog

Coastal Fog

» It forms when the wind pushes warm air off the land over cold seawater.
» It can be extremely dense. Sea kayakers should remain on shore until the fog has lifted.
» It can last all day.

Coastal fog

Steam Fog

- » It forms after a rain shower if the ground is warmer than the rain and looks like steam rising off the ground.
- » It also occurs in the early morning or evening when the air over a lake or river is colder than the water.
- » It is typically short-lived, dissipating by midmorning as the sun warms the air.
- » It feels uncomfortably damp.

Steam fog

Freezing Fog

» It consists of supercooled water droplets,
   which remain liquid even though the
   temperature is below freezing.
» It forms at night.
» It usually means a warm front is pushing out a
   cold front.
» It causes rime to form on the windward side of
   vertical surfaces.

Rime from freezing fog

Valley Fog

» It forms in mountain valleys as a result of a
   temperature inversion.

- » Expect the top of the mountain to be warmer than the trailhead.
- » If conditions are calm, it can last for several days.

Valley fog

Dealing with Fog

- » If you are in the mountains, stay in the trees, or at least on the edge of them, to add contrast in low-visibility conditions.
- » Above tree line follow the rock cairns (stacks of rocks) to remain on the trail.
- » Unless you are competent at navigating with a compass and a GPS, if the fog is so thick that you can't see the trail markers, stay put until it lifts.

- » Stay ashore.
- » Wear rain gear and put a rain cover over your pack to keep yourself and your gear dry.
- » If you are hiking or paddling after dark, use a flashlight instead of a headlamp and hold the beam low to the ground or water for better visibility. A yellow or colored light penetrates fog more effectively than a white light.

Hiker following rock cairns through fog

Predicting Weather by Fog

If the dawn is gray and there is fog in the valley, the weather will likely become nicer as the day goes on. A foggy morning that burns off by noon will usually stay clear for the remainder of the day.

Chapter Five:
Local Effects

Ever find that the weather in town was not even remotely the same at the trailhead? Often local topography overrides the prevailing weather patterns. Knowing how the geological aspects of your surroundings influence weather will help you more accurately plan for it.

MOUNTAINS

Adiabatic cooling is the process by which air cools off as it rises and warms up as it sinks. Air cools by 5.5°F for every 1,000 feet of elevation if there is no moisture in it. Add humidity, and the rate slows to about 3.2° per 1,000 feet. Adiabatic cooling can create rain or snow on a mountain even though the surrounding countryside is dry. For this reason it's always important to carry warm layers and waterproof/breathable outerwear whenever a backcountry pursuit involves a gain in elevation. It will be colder on the summit of a mountain; the question is, how cold? And if there's moisture in the air, you may find yourself hiking in a cloud, rain, or both.

Large mountains create their own weather patterns. Wind flows up the mountain during the day as the air warms up, then down the mountain in the

Approach to Mount Rainier on a clear morning.

evening as it cools off. For this reason clouds form around the peak as the day progresses. Therefore, if you want the best chance for a view from the top, get there early before the clouds develop.

Ever notice on a hike how a breeze picks up in the afternoon? It's the downdraft, and another reason to plan lunch rather than an afternoon

Warm Air

Warm

Cool Air

Warm

Mountain weather patterns

snack on the summit. In other words, the earlier you bag a peak, the more you'll see and the less windy it will be.

Be aware that clouds and strong wind can cancel the topographical weather effects. Clouds insulate, decreasing radiant heat loss from the ground. If a strong front approaches, the high winds associated with the front will overcome the upslope or downslope wind and likely affect the temperature as well.

VALLEYS

Valleys are naturally colder than surrounding hillsides because cold air sinks, thus the name "cold sink" for a concentration of cold air in a low-lying area. The green meadow on a valley floor may look like the ideal place to pitch your tent, but you'll be a lot colder there than on a plateau 500 feet higher. In general, the broader the valley and the steeper its walls, the more drastic the temperature change.

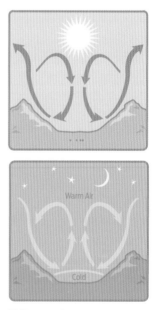

Valley weather patterns

Afternoon shadows on a canyon-bottom campsite.

Interestingly, a narrow canyon radiates heat from one side to the other, keeping the valley floor warmer. That said, it may take longer for the sun's rays to reach the bottom, and they may disappear earlier in the afternoon. The bottom line? You'll be warmer if you camp higher.

LARGE BODIES OF WATER

While all bodies of water increase humidity in the air around them, it takes a huge lake, such as one of the Great Lakes, or a sea to truly impact the weather. The most obvious factors are onshore and offshore breezes along the seashore. Water changes temperature more slowly than land. The ocean can take days, even weeks, to warm up or cool down, whereas the land does it

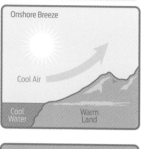

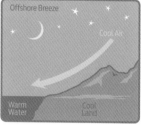

in twenty-four-hour cycles. During the day an onshore breeze blows toward land as air flows from the colder water toward the warmer land. At night an offshore breeze blows toward the water as air flows from the colder land toward the warmer water.

Onshore and offshore breezes are stronger under clear skies and can result in a temperature drop of 10 to 15º. That's why it feels hot in town but cool at the beach. These breezes rarely bring rain. Thick clouds cancel the effect because they prevent a significant temperature differential between the land and the water. Wind on a cloudy day at the coast is due to an approaching front and likely means rain is on the way or perhaps a violent storm.

SNOWFIELDS AND GLACIERS

On mountains where permanent snowfields or glaciers exist, these frozen areas create a downslope breeze that affects about a third of a mile on the hillside below them. This gentle breeze refreshes you on a hot day, but it can make for a chilly campsite. In addition, this cooler microclimate delays the growing season in the affected area. Watch for rare alpine plants that might not grow at the same elevation elsewhere on the mountain.

Trekker by the edge of a glacier.

ALPINE ZONES

Originally named for the climate in the Alps, the alpine zone applies to mountainous areas above tree line yet below permanent snow line. The elevations of various alpine zones differ widely. In New England the alpine zone is generally above 4,000 feet, but there are bald mountains well below that elevation that are exposed to extreme weather and thus support alpine flora. In the northern Rocky Mountains, you have to climb above 8,500 feet—10,000 feet in the middle and southern parts of the range—to get above the timber. While the alpine zone is generally

Alpine zone

dry, with lip-chapping, skin-chilling, dehydrating wind, there may be boggy areas (alpine tundra), and there will certainly be drastic temperature fluctuations between night and day.

Given a choice, it's always warmer and safer to sleep below tree line, but if you must camp on an alpine plateau, pick a small depression to help protect yourself from wind and an unexpected storm. Avoid pitching your tent in an alpine valley, where temperatures will become much more frigid as the downslope wind develops after sunset.

DESERTS

A thermal is a column of rising air that occurs over a hot spot on the land or the water. Giant thermals are common over deserts, where the land gets superheated by the sun. As the air in the thermal rises, it creates a local area of low pressure. The atmosphere seeks to equalize itself, causing air to rush toward the thermal and creating strong winds. A sandstorm can develop as the air rushes horizontally toward the thermal, and a dust devil might also spiral upward. Thermal action builds during the day, making sandstorms and dust devils more likely in the afternoon than in the morning.

Sandstorms occur in all deserts and are among the most violent storms on earth, causing disorientation, difficulty breathing, and extensive damage from

Desert dust devil

the natural sandblasting. Winds can reach 75 miles per hour. It helps to wear goggles, cover your mouth and nose with a bandanna, and cover the rest of your skin with durable wind-resistant clothing until you can find shelter. If you can't find shelter or set up a tent in time, stay low and sit it out to avoid becoming lost. Note: Sandstorms interfere with radio, cell phone, and other electronic transmissions, making them undependable in an emergency.

Chapter Six:
Beat the Chill

While you cannot regulate the atmospheric temperature, you can control your body temperature by what you wear and how you pitch your tent. Getting cold in the backcountry is more serious than in town. Not only is it uncomfortable, it can kill you. Many people die each year from hypothermia, also called exposure.

The weather does not have to be cold for a person to succumb to hypothermia. On a sunny 60°F day at the top of a mountain, you become chilled rather quickly if your shirt is sweaty from your brisk climb. Add a 15-mile-per-hour wind, which accelerates the rate at which heat drains from your body, and you soon start shivering.

Shorts on a frosty mountaintop? These hikers are potential hypothermia victims.

When you become hypothermic, your body cannot sustain a temperature needed for normal metabolic functions. In medical terms, hypothermia progresses in three stages. In stage one your body temperature falls 1.8 to 3.8°F below normal (98.6°). You shiver, have goose bumps, and lose dexterity in your hands, which become numb. You may feel nauseous and tired. In stage two your body temperature drops 3.8 to 7.6°. Though you shiver violently, you feel warm again. You're really not, as evidenced by your pale skin and the blue tint of your lips and extremities. You may also lose coordination, stumble, and seem confused. By stage three your body temperature drops 7.6° or more and your cellular metabolism shuts down. Rather than go into the morbid details, let's prevent hypothermia from setting in.

The key to staying warm is staying dry. It's an old outdoor adage and a true one. "Cotton kills" is another truism. The two go together. Cotton retains moisture and takes a long time to dry out. When wet, cotton sucks heat from your body surprisingly fast, which is okay during the daytime in Death Valley but not atop Pikes Peak. If you wear cotton in the mountains or on water, you risk becoming dangerously cold. You know about dressing in layers; just be sure that all the layers are wool or a technical synthetic fabric. If any of the layers are cotton, the comfortable microclimate that you are trying to create next to your body will disappear.

HOW THE BODY LOSES HEAT

Cause of Heat Loss	Stop It!
Radiation: Heat from your warm skin radiates into the cold air.	Put on more clothing.
Conduction: Heat from your warm skin moves to a cold object when you touch it.	Don't touch. Or, wear a glove or an insulating layer of clothing to block the transfer of heat.
Convection: Cold, moving air carries heat away from your warm skin.	Wear windproof clothing.
Evaporation: Heat from your warm skin is carried into the air as perspiration or other water evaporates.	Wear synthetic garments that aggressively wick moisture away from your skin, keeping you dry. Or wear wool, which retains body heat when wet.

In addition to the fabrics you wear, timing when to put layers on and take them off is critical to regulating your body temperature. You naturally peel off a layer or two when you get warm, but don't wait until you feel chilled to put them back on. In a cold climate carry a down jacket in your pack and throw it on over your other layers every time you stop to rest, *as soon*

WOOL VERSUS SYNTHETICS

Which is the better backcountry fabric, wool or synthetic? It's an ongoing debate. Years ago highly wicking base layers and fleeces of various thicknesses surpassed wool as the backcountry fabrics of choice, but wool has regained favor in recent years. Decide for yourself:

Synthetic fabrics are:

(+/-) Spun from plastic fibers, which might or might not be from recycled materials;

(-) Made from a chemical manufacturing process;

(-) Difficult to recycle once woven into clothing;

(+) Poorer heat conductors than wool. They retain the most body heat of any material;

(+/-) Not moisture absorbent. The fibers wick water away from your body, which feels instantaneous in some fabrics and on a time delay in others;

(+) Quicker drying than other fabrics;

(+/-) Antimicrobial, which helps make them odor resistant, but it's rarely true on day number two;

(-) Easily damaged by fire and bug spray. A spark from a campfire will quickly burn a hole in your fleece pullover, and DEET may damage your favorite T-shirt.

Wool is:

(+/-) Made from a renewable resource, though sheep are not exactly wilderness friendly;

(+) No longer itchy, depending on the quality of the wool and how it was woven;

(+) A poor heat conductor, so it holds body heat in;

(+/-) Able to retain not only a lot of moisture but also body heat, even when wet;

(-) Slow to dry;

(+) Fire resistant. If a spark from a campfire flies your way, it is less likely to leave a hole in your sweater if you quickly brush it off;

(+/-) Less stinky over time, unless it gets wet. Then it smells like wet wool;

(+) Less likely to be damaged by a spark or bug spray.

Hiker properly attired for a cold, wet day

as you stop. It helps to eat a snack, too, which is literally fuel to burn.

The other key to staying warm is covering your head. You lose between 10 and 50 percent of your body heat through your head, depending on whom you ask, but even 10 percent can make a big difference in a cold environment. Put on a wool cap, and you literally cap your body heat in your head.

CAMPING CONSIDERATIONS

Your choice of campsite, tent (and the way you set it up), sleeping bag, and pad goes a long way toward keeping you cozy in the cold.

Picking the Best Tent Site

While there's no such thing as a weatherproof campsite—after all, you are outdoors—where you set up your tent can certainly make it more resistant to wind and precipitation. Select a spot that is on the leeward side of a ridge, and position the door so that it is away from the prevailing wind. Dense trees or a large rock help block the breeze, too. Avoid camping under a solo tree that might attract lightning and low-lying spots that fill with cold air overnight. If your campsite catches the morning rays, the dew or frost will melt more quickly off your tent, and getting out of your sleeping bag will be a less chilling experience. Finally, be sure that your tent site is flat yet won't allow water to accumulate. Pine needles, dead leaves, sand,

A sheltered tentsite that catches the morning sun.

and gravel drain well. Compressed soil and marshy ground do not.

Picking the Best Tent

Most tents are categorized as "three season," "three-season convertible," or "four season."

Three-season tents are perfect for camping in late spring through early fall or in tropical climates, where there is no chance of heavy frost or snow. They typically have lots of netting for a big view of the stars, have a lightweight fly, and are the lightest to carry. They can handle moderate winds.

Three-season convertible tents have a broader seasonal range, from early spring through late fall. They can handle heavy frost and light snow but not heavy snow. They have a moderate amount of netting so you can still get a good view of the stars, but

Four season tents hold up to heavy snowfall.

the walls and ceiling have more square feet of nylon, providing more heat retention. They are heavier to carry but are still reasonable for backpacking. Three-season convertible tents are the most versatile type of tent for most backcountry travelers.

A four-season tent, also called an expedition tent, is for winter camping and for mountaineering, when snow or high wind is a factor. Everything about the tent is heavy duty, from the tent poles to the fabrics, so that the tent can withstand the most extreme weather conditions and keep you the warmest.

Sleeping Bag

Sleeping bags are rated according to the lowest temperature at which they will keep you warm. This is subjective, as some people sleep warmer than others. In general women experience colder extremities (fingers and toes) than men, and people with a higher percentage of body fat stay warmer. To play it safe, select a sleeping bag that's rated at least 10° colder than the coldest temperature at which you'll be camping. In other words, if you expect nighttime lows around 40°, bring a sleeping bag that's rated for 20 to 30°.

The type of insulation inside your sleeping bag is just as important as the temperature rating. If you are camping in a wet environment or if you are a particularly sweaty sleeper, take a bag with synthetic insulation. Down works great until it gets wet.

Sleeping Pad

You can have the warmest sleeping bag on the market, but you'll still freeze if your sleeping pad doesn't insulate you properly from the cold ground. Even in a warm climate, the earth is colder than your body. Without a good pad the ground will conduct your body heat away overnight, leaving you cold and possibly hypothermic.

Sleeping pads should be made of closed-cell foam or be inflatable. Closed-cell pads are heavier and bulkier to pack and less comfortable than inflatable ones; however, if your inflatable pad gets a leak, you are in effect padless. If you opt for an inflatable pad, remember to bring a patch kit. And unless you are trying to save weight while backpacking in mild weather or you don't mind cold toes, a full-length pad is preferable over a three-quarter one.

Pad with a closed-cell foam bottom layer

Canoe loaded with gear

Water-Based Camping

Canoe camping, sea kayaking, and rafting are all wonderful ways to explore backcountry lakes and rivers. It's always chillier on the water than on land during the summer. The water may be 70° compared to the land, which may be 85° or warmer. And there's usually a breeze on the water, for reasons discussed earlier in this book. In addition, you're likely to get splashed now and again no matter how calm the water conditions. Experienced paddlers wear clothing that keeps them dry or at least warm when wet, such as wool or neoprene. It's almost impossible to keep your hands dry around water. Neoprene gloves won't ward off water, but they will help keep your fingers warm.

If the key to staying warm is staying dry, what happens if you fall into water? You will eventually get cold, and rather quickly in most backcountry situations. When water temperatures drop below 70°, here's how long you can expect to survive in the water.

SURVIVAL TIME IN COLD WATER

Water Temperature (°F)	Unconsciousness	Average Survival Time
70–80°	3–12 hours	3 hours–indefinitely
60–70°	2–7 hours	2–40 hours
50–60°	1–2 hours	1–6 hours
40–50°	30–60 minutes	1–3 hours
32.5–40°	15–30 minutes	30–90 minutes
32.5° or colder	Less than 15 minutes	45 minutes or less

The large variations are due to percentage of body fat; conditions in the water, such as salt versus fresh water, which impacts buoyancy; whether you have a flotation device such as a PFD to help conserve energy; overall body mass (smaller people and children die faster because they have a higher surface to weight ratio so they lose heat quicker); and, in some cases, dumb luck.

Chapter Seven:
Cool It

The rules for staying warm in cold weather are the no-no's when it's hot and you want to cool off. Hyperthermia can be just as dangerous as hypothermia in the backcountry. It doesn't have to be 100°F to get overheated. Physical exertion, such as a sustained uphill hike at a fast pace, can cause heatstroke on an 85° day. Most likely you will develop heat cramps and heat exhaustion first, then heatstroke, the most severe form of hyperthermia, which occurs when the body's temperature exceeds 104°. It can be fatal.

In addition to rigorous exercise in high air temperatures, high humidity, direct exposure to the sun, and dehydration can push your internal thermometer dangerously high. The effects of sun exposure are obvious. It heats you up, and your body reacts by perspiring. But dehydration inhibits your body's ability to sweat. If the air is humid, sweat evaporates less readily off your skin, slowing the cooling effect. Any one of these factors alone can cause you to overheat, but if two or more are present, look for early signs of overheating, which include nausea, vomiting, headache, and dizziness if you stand up quickly. In a serious case of heatstroke, you feel confused or hostile; your skin becomes red, hot, and dry; and you may seem drunk before passing out.

STAYING COOLER ON A HOT DAY

Your physical condition, what you wear, what you drink, and how you set up your campsite all contribute to keeping your core cool. It helps to be in shape and not overweight before your camping trip. Your body has to work harder, which generates more heat, moving those extra pounds up the trail. It also helps to be healthy. If you have a fever at the trailhead, your body is already overheated before you take a step. Certain medications, including antidepressants, antihistamines, thyroid hormones, and many "recreational" drugs, increase your chance of heatstroke. If you must take medication, make sure it won't compromise your body's ability to cool itself.

Even if you are fit and healthy, you may still overheat if you're not acclimatized to hot weather. For

Dressing for heat

this reason it's probably not a great idea to plan your first epic backpacking trip of the year during the first hot spell of the spring.

DRESSING FOR HEAT

Wearing summer-weight garments is only the beginning when it comes to outdoor wear in hot weather. Here are some tips to keep you cooler:

» Select loose-fitting clothing of loosely woven fabrics to encourage air movement over your skin.
» Consider covering up with a featherweight long-sleeved shirt and long pants to avoid exposing your skin to the sun.
» Wear a hat that shades your face. Covering your neck, too, is even better.
» Opt for light colors, which reflect sunlight. Black is death in the desert.
» Soak a bandanna in a stream, then put it over your head under your hat or around your neck.
» Soak your T-shirt, then put it back on.
» When you take a break, take off your shoes and socks to cool off your feet. Soaking them in a cold running stream is even better.

Expose feet to the air to cool off.

One way to cool off!

TAKE A BREAK

On a hot day you'll stay a lot cooler if you avoid the sun as much as possible. When you stop for a rest, choose a place in the shade. If there's a breeze, sit so that it blows across you. Consider a siesta, the idea for which originated in tropical climates. If you hike hard early and late in the day, but lay low during the heat of it, you will feel better and reduce the risk of heat-related health issues.

And drink lots of liquids, more than you need to quench your thirst! Put ice in your hydration reservoir

Drink lots of water

or water bottles to keep your water as cold as possible, then drink it all. In desert conditions, where temperatures climb above 100°, a person should consume two gallons of fluid per day. Some of it should be water, but some should also be mixed with juices or sports drinks to replace electrolytes.

What if you are low on water and become stranded? Forget rationing. You'll only succumb to the effects of dehydration more quickly. If you lose just 2.5 percent of your body's weight from water loss (2 quarts of water for a 175-pound man), your physical and mental abilities decrease by 25 percent. The key to survival is keeping your brain hydrated. If you take only a sip now and again or just wet your lips, you are giving water to less important parts of your body, keeping it from reaching your brain. Water has to reach your stomach first in order for your brain to receive it. You're better off if you drink what you need to stay alert and strong, reduce your perspiration rate by staying in the shade, avoid food (especially salty foods and meat) that requires water for digestion, and limit physical activity. There's a saying that you can survive three hours without shelter, three days without water, and three weeks without food. If your survival is at stake, ration sweat, not water. There are countless cases of people dying of dehydration with water left in their water bottles because they tried to ration it. How much water do you need? You need to drink more if your urine is dark or you never have to pee.

WHERE TO CAMP

For the coolest campsite pick a spot in the shade. Contrary to cold weather camping, a valley floor is preferable to a hillside, as cool air sinks. Just make sure it's not dead air. A gentle breeze blowing past your tent will keep you cooler than still air.

Cool, shady campsite

Chapter Eight:
Weathering Wind

Every chapter in this book so far has mentioned wind, why it occurs, and how it affects your comfort outdoors. Wind speed and direction can be helpful tools for predicting weather. Let's begin with wind direction. The following predictions based on wind direction are true most places in the United States:

WIND FROM THE NORTHWEST

» In clear weather expect twenty-four more hours of clear weather and lower temperatures if the barometer is rising.
» In rainy weather expect clearing in several hours and lower temperatures if the barometer is rising.

A windy summit

WIND FROM THE SOUTHWEST

» In clear weather expect twelve to twenty-four hours of continued clear weather if the barometer is rising.

» In stormy weather expect the storm to pass within six hours if the barometer is rising.

» In clear weather expect rain in the next twelve hours if the barometer is falling.

» In rainy weather expect heavier rain, then clearing after twelve hours if the barometer is falling.

WIND FROM THE SOUTHEAST

» In clear weather expect continued fair weather if the barometer is rising.

» In stormy weather expect clearing very soon if the barometer is rising.

» In clear weather expect rain within twelve hours and high winds if the barometer is falling.

» In stormy weather the storm will get worse, then clear out in about twenty-four hours if the barometer is falling.

WIND FROM THE NORTHEAST

» In clear weather expect continued fair but cooler weather if the barometer is rising.

» In stormy weather expect clear, cool weather to arrive soon if the barometer is rising.

Wind from the northeast can bring heavy snow.

» In clear weather expect stormy weather in twelve to twenty-four hours if the barometer is falling.

» In stormy weather expect very heavy precipitation, gale-force wind, and much colder temperatures if the barometer is falling. In the northeastern United States, this weather pattern is called a "nor'easter."

Note: You do not need a barometer to tell barometric pressure. If you have an altimeter, you can get a sense of the trend if your altimeter shows a change in elevation, even though you haven't moved. If your altimeter shows a rise in elevation, it really means the barometric pressure has fallen and a low-pressure system has arrived. A fall in elevation indicates a rise in barometric pressure or an incoming high-pressure system.

WIND SPEED

Wind speed and direction are less complicated than barometric pressure. If the wind is picking up, a front is approaching. If it's a south wind (that is, from the south), it's going to rain or snow. If it's from the north or west, it's going to be clear and probably colder. Winds rarely come from the east in the United States, as our prevailing winds are westerly. If they do come from an easterly direction, it's likely from the southeast or northeast. Increasing wind from the northeast means it's going to get really nasty very soon and it could last a while.

The other key aspect of wind speed is windchill. Wind does not lower air temperature. It just feels colder because wind draws heat away from exposed skin. The stronger the wind is, the faster the heat loss. Cover up, and you effectively cancel the effects of windchill. The chart from the National Weather Service on the facing page shows the windchill at a particular air temperature and wind speed. The chart is color coded to tell you how quickly you would develop frostbite on exposed skin. (Go to www.nws .noaa.gov/os/windchill/index.shtml.)

Chinook winds. Chinook winds, also called "foehn winds" by meteorologists, are strong, steady winds usually 20 to 25 miles per hour, though they can blow much harder. They occur around mountains in the western United States and cause a dramatic

Temperature(°F) \ Wind Speed(mph)	Calm 5	10	15	20	25	30	35	40	45	50	55	60
40	36	34	32	30	29	28	28	27	26	26	25	25
35	31	27	25	24	23	22	21	20	19	19	18	17
30	25	21	19	17	16	15	14	13	12	12	11	10
25	19	15	13	11	9	8	7	6	5	4	4	3
20	13	9	6	4	3	1	0	-1	-2	-3	-3	-4
15	7	3	0	-2	-4	-5	-7	-8	-9	-10	-11	-11
10	1	-4	-7	-9	-11	-12	-14	-15	-16	-17	-18	-19
5	-5	-10	-13	-15	-17	-19	-21	-22	-23	-24	-25	-26
0	-11	-16	-19	-22	-24	-26	-27	-29	-30	-31	-32	-33
-5	-16	-22	-26	-29	-31	-33	-34	-36	-37	-38	-39	-40
-10	-22	-28	-32	-35	-37	-39	-41	-43	-44	-45	-46	-48
-15	-28	-34	-39	-42	-44	-46	-48	-50	-51	-52	-54	-55
-20	-34	-40	-45	-48	-51	-53	-55	-57	-58	-60	-61	-62
-25	-40	-47	-51	-55	-58	-60	-62	-64	-65	-67	-68	-69
-30	-46	-52	-58	-61	-64	-67	-69	-71	-72	-74	-75	-76
-35	-52	-57	-64	-68	-71	-73	-76	-78	-79	-81	-82	-84
-40	-57	-63	-71	-74	-78	-80	-82	-84	-86	-88	-89	-91
-45	-63	-72	-77	-81	-84	-87	-89	-91	-93	-95	-97	-98

Frostbite Times on exposed skin: 30 minutes / 10 minutes / 5 minutes

temperature increase, sometimes 40°F in less than an hour, which is why Chinooks are sometimes called "snow eaters." They blow for a long time, often over twenty-four hours.

Valley winds. Valley winds flow up or down a valley floor depending on the time of day and length of the valley. If you're planning to paddle or raft through a canyon or long valley, know that you might encounter a headwind later in the day, as warm air from the end of the valley blows into it. A valley wind always blows upriver, into the valley from lower land, during the day, then downriver at night.

Deep canyons are susceptible to valley winds.

Chapter Nine:

Extreme Weather

Extreme weather includes violent thunderstorms, hailstorms, and blizzards, which might be considered normal depending on your location. It also includes hurricanes and tornadoes. These violent weather systems are inherently dangerous when you are at home, let alone in the backcountry. However, you can improve your chances of survival and perhaps get through the storm unscathed if you know what to do.

THUNDER AND LIGHTNING

When there is thunder, there is always lightning, even if you can't see it. As an updraft, or fast-rising column of air, cools and condenses, an equally aggressive downdraft, usually filled with heavy rain, occurs. You can feel the downdraft ahead of the storm, not only because you're getting wet but also by the rush of cold air. You can see the location of the updraft, 2 to 3 miles away, by the towering anvil-shaped cumulonimbus clouds. When you notice these weather conditions developing, get below tree line immediately, and take cover in a cluster of trees. Lightning is close at hand! If you are in a boat, go quickly to shore. If it's too far, stow your paddle or oars parallel to the water (not sticking up) and get as low as possible in the middle of the boat, avoiding contact with wet objects. Regardless of your location, turn off all electronic devices, such as cell phones and GPS units, which make good electrical conductors.

Lightning is caused by friction. As minute particles churn skyward in the updraft and toward the ground in the downdraft, they create a strong electrical charge—lightning—between parts of the cloud or between the cloud and the earth. A single lightning bolt discharges up to 30 million volts of electricity, explosively heating and compressing the air, which you hear as thunder.

Tree scarred by lightning.

You see the lightning before you hear the thunder, as the flash of light from a lightning bolt travels at the speed of light (186,000 miles per second), whereas thunder travels much slower, at the speed of sound (1,125 feet per second). If you count the seconds between the flash of lightning and the clap of thunder, then divide by 5, you can determine how many miles away the storm is. The speed at which the storm reaches you depends on the wind.

Lightning in the Alpine Zone

If a thunderstorm catches you on an exposed ridge or mountaintop and you cannot get below tree line, do the following:

» Get off the top of the ridge or the summit, even if it's a just few feet lower, so you are not at the highest point.
» Get away from your pack. The metal in it attracts electricity.
» Crouch on a dry surface, preferably on your sleeping pad, if it's dry, to insulate you from the ground. Water conducts electricity.
» Do not lie down or sit. Crouch on the balls of your feet, arms by your sides, to minimize your contact with the ground and thus your susceptibility to ground current from a nearby lightning strike.
» Avoid single trees, which act like lightning rods.

- » Spread out your group, leaving at least 25 feet between each person to reduce the chance of everyone being struck by lightning.
- » Avoid wet depressions or caves. The moisture that collects there conducts electricity.

TORNADOES

Arguably the most intense of all storms, tornadoes are defined as violently rotating columns of air that extend down from a thunderhead to the ground. If one touches down over water, it is called a "water spout."

As a thunderstorm forms, the change in wind direction and increase in wind speed create a horizontal spinning effect in the lower atmosphere. The strong updraft associated with a cumulonimbus cloud tilts that spinning air system upward. At first it appears in the sky below the thunderhead as a separate "wall cloud." Typically, there is heavy rain around the wall cloud, but little or no rain under it. If the rotating air associated with the wall cloud finds its way to the ground, it becomes a tornado, the center of which contains an area of extreme low pressure. Air rushes in to fill the void, which can result in horizontal winds in excess of 300 miles per hour and vertical winds of 150 miles per hour.

Flying debris causes most tornado-related injuries and deaths. If a tornado develops near you in the backcountry, do not try to outrun it. You can't, but you can take the following steps:

» Curl up in a tight ball in a ditch or other deep depression, but be alert for potential flooding.
» Cover your head.
» Do not get into a vehicle.
» Avoid taking shelter under trees, particularly conifers or other trees with shallow root systems.
» If near a lean-to, curl up against the outside wall of it and cover your head.
» Do not get into your tent. You are in more danger inside a tent, or any other structure for that matter, than you are outside it.

HURRICANES

A hurricane is defined as a severe tropical cyclone with winds greater than 74 miles per hour. Hurricanes originate over the Atlantic Ocean, the Caribbean Sea, and the eastern Pacific Ocean. They travel north, northwest, or northeast and bring heavy rain as a result of moisture extracted from the sea en route to land. Unless you are in a coastal area, hurricanes affect backcountry travelers mainly with impressive downpours, although the wind can be strong, too.

Hurricane season runs from June 1 through November 30, though most develop from early

August through mid-October. It is difficult to anticipate the approach of a hurricane. Your best bet is to check the National Weather Service forecast before you leave home. If a tropical depression is forming or if it has already developed into a hurricane, check its likely path. Delay your trip if you will be in that path. Otherwise bring good raingear, and plan to spend extra time in your tent. And remember that a hurricane that hits land in North Carolina can still cause heavy rain in New England.

SLEET AND FREEZING RAIN

Sleet and freezing rain are similar. Sleet looks white and freezes before it reaches the ground. It bounces when it hits. Freezing rain turns to ice upon contact with the ground or an object. It's clear and coats things.

Freezing rain usually falls first, at the leading edge of a warm front. Sleet follows on the north end of the freezing rain line. Sleet and freezing rain occur when there are several layers of warm and cold air at or just above freezing level, such as when a cold air mass arrives at the same time as a storm system. Under these conditions precipitation freezes, melts, and refreezes en route to the ground.

Sleet and freezing rain are not particularly extreme except for the slippery footing they create; however, if they turn into an ice storm, which coats tree limbs to the point at which they sag and break, it's a hazardous time to travel in the backcountry. A three-season tent will likely collapse under the weight of the ice. If an ice storm develops when you are in the mountains, go to a trailhead immediately, and then get indoors. If you cannot get out of the woods, look for a low rock overhang or similar spot for shelter. If there's enough snow on the ground, build a snow shelter or dig a snow cave. Even a partial shelter or cave will help protect you.

HAIL

Hail is different from sleet. Sleet is a type of winter precipitation, whereas hail can fall any time of the year. Sleet is fine like snow. Hailstones can be as small as a BB or as large as a grapefruit. Hail forms in conjunction with thunderstorms and tornadoes in cumulonimbus clouds. Hail does not always accompany a severe thunderstorm; however, if it is present, it indicates extreme weather. Large hailstones are like falling rocks; if you have a climbing helmet, put it on. At least cover your head with a hat, any hat, then take shelter under the nearest thick evergreen. Hail falls from above, so your best protection is to put something sturdy between you and the sky. Though the branches of a spruce tree aren't a solid roof, the many layers of needles help deflect falling hailstones.

Stay put in blizzards.

BLIZZARDS

A blizzard is a snowstorm with winds over 35 miles per hour and visibility of less than one-quarter mile. The chance of hypothermia is extremely high, not only from cold temperatures but also strong wind. Breaking trail in a blizzard, whether on foot, snowshoes, or skis, is exhausting work, and it is nearly impossible to navigate even on a well-defined path.

If you are winter camping and a blizzard strikes, plan to stay put in your tent until it passes. A decent four-season tent can withstand a fair amount of falling or drifting snow if it is properly set up. Monitor the snowfall against the tent, and clear it away as you need to. Also make sure that your tent is well ventilated. If you don't have a tent, you should build a snow shelter or a snow cave to protect yourself from the elements.

HOW TO BUILD A SNOW SHELTER

Building a snow shelter, also called a "quinzee," takes time, three to four hours, but it can save your life in a blizzard that lasts all day. You'll stay warmer from the exertion and then have a place in which to rest and wait out the storm.

1. Clear an area of snow 7 to 8 feet across.

2. Make a large dome of snow about 6 feet high. Mixing snow from different depths in the snowpack encourages the hardening process, called "sintering."

3. Insert several sticks at least a foot in length into the top of the dome.

4. Allow the mound to set up for one to three hours, until it is hard enough to dig through without collapsing.

5. Create an entrance that's low and straight into the dome.

6. Dig upward at an angle, eventually carving out an elevated sleeping shelf. The sleeping shelf needs to be higher than the floor so that cold air will sink and depart through the door.

7. Make a windbreak in front of the door, though you can use your pack as the door itself.

Quinzee snow shelter

8. Hollow out the rest of the interior, but be sure to leave at least a foot of thickness all around. If you come to the interior end of a stick from Step 3, you've cleared the maximum snow in that spot.

9. Pull out a couple of larger sticks from Step 3, or poke a ski pole through the roof to create a vent hole.

FLASH FLOODS

Flash floods occur literally in a flash and are extremely dangerous. According to NOAA, more people have died in flash floods over the last thirty years than in hurricanes, tornadoes, blizzards, or lightning storms. They are typically caused by a heavy downpour, at least an inch per hour, in an area where the ground is either too hard or too saturated to absorb precipitation. The southwestern region of the United States is particularly susceptible to flash floods due to predominantly desert conditions and an abundance of slot canyons. Even if the downpour is miles away, it can flow to you. For this reason, it's important to obey the maxim, "Turn around, don't drown." In other words, rather than crossing a flash flood area, go around it, even if it's a sunny day. Save your trek through places like Bryce Canyon for a time when a stable high pressure system is over the region. If you accidentally hike into an area that might be susceptible to flash flooding, such as a dry riverbed in a deep valley, move through it quickly and try to stay on higher ground.

Chapter Ten
Lore or Likely

Humans have tried to forecast weather since the beginning of time. Lacking Doppler radar and satellite surveillance, they've read the sky, listened to animals, watched plants, and felt the wind for clues to incoming weather. Some old wives' tales are surprisingly accurate. Others get lucky now and again. Let's sort out nature's reliable weather predictors from what's strictly lore:

Judge the distance of an approaching thunderstorm by counting.

True. It is possible to judge the distance of an approaching thunderstorm by the time lapse between the thunder boom and the flash of lightning. Count slowly, "One one-thousand, two one-thousand, three one-thousand . . ." so that each "one-thousand" equals approximately one second. For every five seconds, the storm is one mile away. For example, if you count to "three one-thousand," the storm is just over half a mile from you. Better hunker down!

Tornadoes never occur in the mountains.

False. No place is safe from a tornado. Though less frequent in mountainous areas, they do happen, even at high elevations. In 2004 a tornado touched down in Sequoia National Park at an elevation of 12,000 feet.

Red sky at night, sailor's delight

The color of the sky at sunrise and sunset tells the incoming weather.

True and false. You've heard the saying, "Red sky at night, sailor's delight. Red sky in the morning, sailors take warning." The redness of the sky is caused by the sun's rays reflecting off dust particles when there's little or no cloud cover. As our weather typically comes from the west, a red sky as the sun sets in the west means a high-pressure system (good weather) with stable air is coming. A red sky as the sun rises in the east means the high-pressure system has passed to the east. A storm system may be approaching, especially if the sky is deep, fiery red, a sign that it contains a lot of water vapor. However, if the air is polluted, all bets are off. Air pollution can cause red skies both morning and evening regardless of the weather pattern.

Geese won't fly before a storm.

True. Perhaps geese, which are large, heavy birds, have a harder time taking off when air pressure is low because the air is slightly thinner. More likely

they simply sense the impending storm and stay put. Other waterfowl do the same, as do seagulls, which cluster on a sheltered beach.

If the bubbles in your coffee gather in the center of your cup, the day will be fair.

True. Java forecasting relies on the way air pressure affects the surface tension of coffee in a mug. Note: It has to be strong, brewed coffee to have enough oil in it to work, and the mug must have straight up-and-down sides. Stir your coffee, creating bubbles. If the bubbles amass in the middle, you're in a high-pressure system, which is making the surface of your coffee slightly convex (higher in the middle). The reason? Bubbles are mostly air, so they migrate to the highest point. It's going to be a beautiful day. If the bubbles form a ring around the sides of the mug, you're in a low-pressure system, making the surface of your coffee slightly concave (lower in the middle). Rain is likely.

Bugs disappear an hour before a storm.

True. Mosquitoes and blackflies are thickest from about twelve hours to one hour before a storm hits,

then they take cover. If the bugs stop biting, put on your raingear. Bees stay close to the hive when the weather is going to get worse.

Springs flow faster when a storm approaches.

True. Low barometric pressure associated with an approaching storm will cause a natural spring to flow out of the ground faster. The same low pressure will also cause ponds to look cloudier as more muck from the bottom rises toward the surface due to a higher volume of marsh gases.

Caves give off cold air when a storm approaches.

True and false. If you stand in front of a cave when a low-pressure system approaches, you feel cold air rush out. However, this is not the only time a cave "breathes out." If the temperature outside the cave is warmer than the temperature inside the cave, the cold air will rush out as it naturally attempts to equalize with the warm air.

Hair gets curlier as humidity rises.

True. It depends on your hair, but for many people, their hair gets wavier or curlier as humidity rises because hair tends to contract (curl) when it's wet and relax (straighten) when it's dry. Canvas, hemp, and other natural fibers do as well. However, wood does the opposite, expanding when wet and contracting when dry. If your axe has a wooden handle, the handle feels looser in fair weather and tighter if rain is coming. The higher humidity preceding the storm swells the wood.

Songbirds sing louder just before a storm.

False. Some people believe in the opposite theory, that songbirds become quiet just before a storm. It depends on the species. Unless you really know your birds, this is an unreliable method of weather forecasting.

Sound travels farther when a storm approaches.

True. High humidity and increased wind from the approaching low-pressure system both help carry sound waves farther.

If the wind dies suddenly, it's about to pour.

True. Often called the calm before the storm, if the wind has blown steadily for the last few hours and clouds are developing, then suddenly it all stops, take cover. You've got only moments before it pours. Don't confuse this phenomenon with the "eye of the storm," a calm period during a hurricane when the center of the hurricane passes over you. If you are caught in a hurricane, stay put at the first sign of calm weather. You've only experienced half the fury.

If campfire smoke tonight rises in a straight column, expect a fair day tomorrow.

True. On a calm night smoke rising vertically means a high-pressure system is upon you. On the other hand,

Rising campfire smoke means a fair day tomorrow.

if campfire smoke stays low to the ground, then disperses, a low-pressure system has arrived, and you could get wet.

Tell the temperature by counting cricket chirps.

True. If you count the chirps of a cricket for fourteen seconds, then add forty, you can figure out the air temperature (in °F). For example, if you hear twenty chirps, it's 60° outside. Crickets are correct within a degree or two more than 75 percent of the time. Close enough.

Deer migrate to a lower elevation when the weather is about to worsen.

True. All hoofed animals, such as deer, bighorn sheep, elk, and moose, head for sheltered valleys when a storm approaches. If you notice ungulates in the lowlands, hopefully you are already wearing your raincoat. The storm is virtually upon you.

If you see a circle around the moon, it will rain or snow soon.

True. The moon can be an excellent weather indicator. If the moon is clear and bright, a low-pressure system has cleared the dust out of the air, so expect rain. A halo around the moon, which is caused by cirrostratus clouds, means a warm front approaches. Expect precipitation, but it might take two to three days to arrive.

RAINBOWS

Nature has many ways of hinting about the weather, but nothing is surrounded by as much lore as the colorful rainbow:

When looking at your natural surroundings for hints to the weather, your chances of an accurate forecast will be much higher if you observe as many indicators as you can, then make your prediction. Always remember that weather forecasting, even with the most advanced technologies, is not an exact science, though you can get darn close to an accurate forecast if you pay attention to the clues around you.

Rainbows are good weather predictors.

What you see:	Forecast:
Rainbow in the east in the evening	It's going to rain.
Rainbow in the west in the morning	It's going to rain.
Rainbow at lunchtime	A sudden downpour is about to hit.
A broken rainbow on a cloudy day	Blustery, stormy weather approaches.
Quickly fading rainbow	Fair weather approaches.
Rainbow in the afternoon	Good weather is coming.
Rainbow over water, but not touching it	Clear weather approaches.
Rainbow visible from a long distance	Clear weather approaches.
Rainbow disappears suddenly	Fair weather approaches.
Double or triple rainbow	It's fair now, but heavy rain will arrive soon.

INDEX